The Little Book of Big Questions

A Coffee Table Book for Thinkers and Questioners

J Edward Neill

Cover Graphics and Interior Art by J Edward Neill

Tessera Guild Publishing

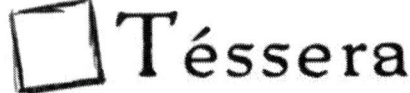

Copyright © 2015 J Edward Neill

All rights reserved.

ISBN- 13: 978-1537066141
ISBN- 10: 1537066145

I can't get enough.

So many questions exist in the universe, I could spend my whole life asking and not even crack the surface of everything there is to know.

So what else is there to do…

…but ask anyway?

Thank you for reading The Little Book of Big Questions. This book is a compilation of my two favorite volumes in the Coffee Table Philosophy series.

101 Deeper, Darker Questions for Humanity contains imaginative, mind-stretching inquiries. Some are far-fetched. Others reach into the darkest part of humanity's soul. The topics are human relationships, philosophy, good vs evil, and night vs day.

101 Questions for the End of the World is *truthier*. It approaches science, religion, and philosophy from an ultimate 'I don't know' angle. If you're able to answer these without scratching your head at least a little bit, congratulations – you're smarter than most people. Seriously.

Enjoy…

Contents

Part I – 101 Deeper, Darker Questions for Humanity

Part II – 101 Questions for the End of the World

101 Deeper, Darker Questions for Humanity

If you could hide in a dark corner and ask the universe questions, would you?

...of course you would.

Six Deadly Sins

The 7 Deadly Sins are:

Envy

Greed

Sloth

Lust

Gluttony

Pride

Wrath

If you could destroy one of these *forever*, as in remove it from the consciousness of every human being for all time, which sin would you choose?

First Snowflake in the Avalanche

In your opinion, what is the *smallest* act by an enemy (against your home nation) that you would consider worthy of starting a full-scale war?

Anything by The Cure

Name at least one song you like that makes you sad whenever you listen to it.

And explain why you like it.

The Time Haters

Suppose someone truly evil comes to power during your lifetime.
This awful person enslaves thousands of people, steals billions of dollars, and kills whomever they want.
...whenever they want to.
Now suppose you have a time machine that can take you back exactly 100 years into the past.
Would you ever consider seeking out the evil person's ancestors and *removing them from the equation*?
If so, what would your method be?

The Unfriending

Which of the following *one-time occurrences* might cause you to lose your faith in a good friend?

- They fail to show up for a social event and never explain why
- You're forced to pay a large restaurant tab because they didn't bring any money
- While tipsy, they flirt with your significant other
- They borrow something and 'forget' to return it
- They send terrible gifts for birthdays and holidays
- They tell you 'white-lies' to spare your feelings

Rank and File

In your opinion, is any one branch of the military *braver* than the others?

For example, are Marines braver than fighter pilots?

Do infantry soldiers have more courage than the crew of a submarine?

Forever Off the Grid

Suppose a person became extremely wealthy early in life. By the time they're 30 years old, they've got billions of dollars in the bank.
Now suppose this person retires from any and all forms of work.
They don't have a job.
They don't do volunteer service.
They don't work for any charities.
They'll never have any children.
Is this person a bad human being?
Are they lazy? A total non-factor in society?
Or is it absolutely fine for someone to be this way?

Gandalf the *Black*?

Suppose magic is real and every person can learn one specific power to claim as their own.

What power will you wield?

What will your 'wizard' name be?

The Chisel

You died yesterday.

Your family and friends commissioned a tombstone to forever mark your grave.

In terms of the life you've actually lived, what saying appears on your tombstone?

Suppose you'd lived ten years longer than yesterday.

Given the extra time, what then would your tomb say?

And if an enemy or rival of yours were to engrave your tomb, what would they write?

Life Long or Die Hard

In Shakespeare's play, *Julius Caesar*, the following line is uttered:

"Cowards die many times before their deaths;
The valiant never taste of death but once."

In other words, Shakespeare means to say that those who live in fear die a small death every time they back away from something that terrifies them.

Do you agree with this?

Why or why not?

Honesty Setting: 90%

Among most humans, it's assumed not everyone is completely honest 100% of the time.

Some people exaggerate.

Other people understate.

Others leave out details so as not to offend whoever's listening.

If you could choose an honesty setting for all the people in your life, what % would you pick?

And what's your personal honesty setting?

A Day of Infamy

Suppose everyone alive knew the exact date of their death.

Not the method or cause, *just the date*.

If your death were scheduled to be twenty years from now, would you live any differently than you currently do?

What if your death were only *five* years away?

If for either situation you said you'd change your way of life, shouldn't you change it now anyway? ☺

Is there any one part of your day that is absolutely sacred to you?

As in a small daily moment, a place in time, an everyday activity, or a meal?

Also...

Is there any part of your daily life you dread? As in; you'd live happier without it?

Heart of Darkness

And now, in ten words or fewer, state your personal definition of HATE.

Behind the Veil

Throughout history, many millions of people have reported seeing ghosts, apparitions, aliens, monsters, and other strange, unexplainable phenomena.

Which of the following do *you* believe probably exist?

Ghosts

Aliens

Angels

Demons

ESP (extra-sensory perception)

Alternate Dimensions

Got any proof? ☺

If there is such a thing...

In ten words or fewer, state your personal definition of LOVE.

Silver Tongues

What is the worst lie you've ever told?

If you're reluctant to answer, then...

...what is the worst lie someone has ever told you?

And why?

Peak and Valley

For you personally, what is the absolute best part about sex?

And...

What's the hardest part about sleeping with a new person?
(no pun intended)

Mic Drop

Other than winning the lottery, name three things that would make you want to immediately walk away from your job.

As in; tomorrow morning you pack up your things and never go back.

Calculus

Fill in the blanks. No more than two words each.

_____

$+$

_____

$=$

the best reasons for living.

Hard Scales of Justice

Imagine the following scenarios.
In which of these (if any) would you support the *death penalty*?

- After fifty years of marriage, an elderly woman shoots her husband in his sleep. Her motivation is to claim life insurance money
- A pair of eleven-year old boys lure a little girl into the woods, where they murder her for no apparent reason
- A man foreign to your native country guns down ten people in a small, peaceful café
- A politician orders your military to bomb a village in another country, killing *one* enemy combatant and *two-hundred* civilians

Automatica

Do you think humanity will ever reach a point of not having to work?
As in; *everything* will be automated.
Labor won't be needed.
Everyone will have unlimited leisure time.

Is this something humanity should try to achieve?

Is it a future you'd embrace?

Dying Stars and Sabertooth Tigers

Name something that is both *beautiful* AND *terrifying* to you.

Venus & Mars

This is an exercise for couples.

...or even platonic friends, so long as one is a guy and the other is a girl.

Give each guy and girl a pen and one sheet of paper.

With a 60-second time limit, everyone writes down their personal answer to:

What five things are needed to make a perfect date?

And then everyone shares.

How different are the answers between the sexes?

Little Bang Theory

Some people have theories about how the world began.

And how it will end.

And maybe even theories about what it all means.

But...

What if you— *yes you*—could decide how it all began, what it all means, and how you'd like it to end?

Play god for a moment:

How would *you* like for the universe to have begun?

How do *you* want it to end?

What do *you* want it all to mean?

The Last Bottle's Bottom

It's entirely possible humanity will never develop the ability to leave Earth and spread out to live among the stars.

If so, how do you imagine the *very last human* will die?

Picket Fences

You're *madly* in love.
Tomorrow you'll have the chance to marry the love of your life.
You'll have a huge house full of beautiful things.
Your children will be smart and loving.
On the surface, your marriage will appear ideal.
But here's the thing: You've glimpsed the future and have seen that while your marriage will be stable and polite, it will ultimately become passionless and empty. Knowing what you know, are you still walking down the aisle tomorrow?

Angel of Death

You *lucky* bastard.

Or maybe not.

You've just acquired a new ability.

From now on, you can wish anyone in the world dead.

If you use this power, not only will the person die instantly and painlessly, but you'll also gain a million dollars for each person you use it on.

How many times (if any) do you think you'll use this power?

On whom?

What would you do with all that money?

Can we Talk?

Take a brief trip into your own past.

What's one piece of advice you'd give your younger self?

Idealistic

For the following situations, state whether you believe each one is *society's* job or the job of the individual *family*:

- Extending the life of a terminally ill patient
- Doing the same ↑ for someone who is a war veteran
- Providing treatment for a criminally insane person
- Providing a free college education
- Providing welfare for underprivileged children
- Caring for the elderly

Little Pharma

Turns out you're in the pharmaceutical industry.
And you've just invented the pill of a lifetime.
Used once daily, whoever takes it becomes immune to _____.

_____ can be whatever you want it to be, but it can only be *one* thing.
A disease. A mental condition. A state of mind.
Whatever.
Fill in _____.

Extinction Level Events

Choose one animal species on Earth to utterly eliminate. Consider all the repercussions of their absence.

Now choose an extinct species to revive and return to a healthy population level.

Explain both your choices.

Protect this House

Name two instances in which it's ok to be completely selfish.

The Heist

If you could steal any *one* thing in the world and make it yours forever, what would it be?
It can be an object, a person, a life situation, a place.
You won't get in any trouble for taking it.
No one will ever know.

Well?

Monopoly Money

You've just received three GET-OUT-OF-*DEATH*-FREE cards.
Anyone who has one of these avoids the next time they would die.
Once a death is avoided, the card vanishes.
So...
Keeping all three for yourself?
Giving any away?
Distribute your cards and explain.

Fix-it Felix

What is humanity's biggest flaw?

i.e.; What's the one weakness we all possess that you'd like to put an end to.

Got any suggestions for fixing it?

Soul Mining

Imagine you have no work to do.
You have a free home and vehicle.
All your food is prepared for you at no cost.
100% of your time is reserved for leisure.
Even if you wanted to get a job, you couldn't. There are *none*.
The catch: you have no money for expensive vacations, travel, or lavish entertainment.

What will you do with all your free time?

9 Billion Vampires

Suppose 100 years from now humanity perfects a sustainable scientific method for extending human life.
Indefinitely.
Meaning everyone everywhere receives an annual treatment that effectively makes them immortal. No diseases. No aging. No natural death.
What are some of the problems you foresee should this ever happen?
Design two laws you think would be needed to maintain peace and prosperity among an immortal population.

Circuits Crossed

Can feminism and chivalry co-exist?

As in; pure, genuine chivalry and sharp, dedicated-to-equality feminism?

Explain.

Nectar of the Gods

Suppose you had a single bottle of the most delicious drink ever to exist.
It's the perfect beverage, smooth as silk, subtle and intoxicating, and utterly refreshing.
There's only one bottle of this nectar left in the world.
So…
What occasion would you use to drink it?
Would you finish it off in one night or spread it out over a longer period?
Would you share it with anyone?

The Vow

Suppose right here and now you are compelled to make an *oath*.
Meaning you'll kneel on the ground, say some sacred stuff, and commit to fulfilling this oath for as long as you shall live.
It can be anything you want, but it must require a lifetime of obligation.

What is your life-oath?

Metaphysical Exam

Of the following qualities, pick the two you'd most want as traits of a close friend.

Brutally honest

Entirely non-judgmental

Rarely complains

Never envious or jealous

Extremely creative

Always the fun one

Highly responsible

Now choose again (just two) to be traits of your lover or spouse.

Any difference?

Ctrl, Alt, *Delete*

What's one email, text, or phone call you wish you could go back and *undo*?

On a Scale of 0-10...

...in which 0 is 'not at all', 5 is average, and 10 means

'*highly*':

How intelligent are you?

How physically attractive are you?

How charming?

How artistic?

How generous?

And how narcissistic?

Needs of the Many

Imagine you and *ninety-nine* other people are fleeing through the wilderness.

A vicious, unstoppable beast pursues you, but for the moment has lost your trail.

The only problem: an infant child in your group has begun to wail, howl, and cry uncontrollably. If it continues, the beast will surely hear the noise and devour all one-hundred of you.

So...

The baby isn't going to stop crying anytime soon. What do you do?

Population *Destiny*

At the time this book was written, the approximate population of Earth is 7,200,000,000.

That's 7.2 billion human beings.

Now suppose tomorrow you wake up and Earth's population is only 10% of this.

As in only 720,000,000 – distributed evenly across the globe.

Would everyone's life be *better*?

Worse?

What do you think it'd be like to live on Earth with a population density so low compared to now?

Mirror, Mirror, on the Wall

In your own words, define what you believe the difference is between a terrorist and a normal soldier.

Swords and Cajones

Imagine everything in the world is exactly as it is right now.

Except...

No modern weapons exist. No guns, missiles, bombs, warplanes, tanks, nukes, etc.

Instead, war is fought with swords, spears, axes, and shields.

It's *medieval.*

Would there still be just as much warfare?

Daggers for the Sky

You've been assigned a monumental task.

Literally.

Over the next twenty years, you'll be the project leader for ten- thousand workers as they construct a monument to represent all of humanity.

You have a one-billion dollar budget.

You have total creative control.

What do you instruct your workers to create?

Getting Around

From the following scenarios, choose the most appealing to you:

- Be happily married to one person for your entire life. Never once have an affair
- Be happily married for twenty happy years, but also enjoy a twenty-year period of single-dom, during which you have sex with at least ten different partners
- Never be married. Have as much freedom and as many different sex partners as you desire
- Insert a different answer here _____

Unshakeable

Consider your core beliefs.
Concentrate on the ones regarding the Earth, your fellow humans, and your spirituality.
Is there one belief among these that you don't think you could ever be swayed on?
If so, explain why.

Shopping Lists for Gods

You've just woken up and found one billion dollars in your account.

The money is legit. It's all yours.

Using percentage values between 1-100%, define what percent of your money you'd use for:

Savings

Entertainment

Housing

Vehicles/Transportation

Charity

Gifts to family

Your %'s must total 100.

Hot for Teacher

You're going to be a teacher for one month.

Your class: *every five-year old child in the entire world.*

Given the chance, what lesson would you teach these kids?

What if they were all fifteen-year olds?

180 Degrees

Suppose a man dedicated the first twenty years of his life to being a vicious criminal.
He was a thief, a thug, an arsonist, a kidnapper, and even a *murderer*.
But then, for the next 50 years, he turned his life around.
He gave millions to charity. He found homes for orphans.
He fed the poor. He traveled to war-torn nations and helped innocent people evacuate.
What is the value of this man's life?
In your eyes, has he found redemption?

Bank of Humanity

Do we as human beings have any responsibility to:
Help the homeless?
Put our lives at risk to save fellow humans in danger?
Raise our children to respect cultural norms?
Honor the traditions of the nation in which we live?
Look out for our neighbors?

Scientists have created a new and powerful type of remote control.

With it, you can choose something in the world to turn on or off.

—At any time. From any place. No matter the distance.—

How it works:

You pre-program the remote to turn on or off one *specific* machine, electronic device, vehicle, emotion, or even a person.

This pre-programming works only on one thing. Ever.

What's your remote *control*?

_____ wanted

You've just started a new company. You make, market, and sell stuff all around the world. Which of the following people are you most likely to hire?

- *A beautiful woman fresh out of college*
- *A sharply intelligent, but socially awkward young man*
- *An older veteran skilled in the ins and outs of business*
- *A tireless, hard-working middle-aged man without a degree*
- *A middle-aged woman who excels at team-building*

Explain your choice.

Destitution

Finish the following sentences:

The *poorest* person alive is he who _____.

The *richest* person is she who _____.

Generations of Evil

In certain cultures around the world, different generations are referred to *separately*.

For example, in America there exist such divisions as Gen-X, Baby Boomers, and The Greatest Generation.

It's a common theme for older generations to criticize those who are younger, often with cries of, "Kids these days don't know a damn thing!"

Is it true that previous generations contain people who are wiser, harder working, and more moral?

Or has every generation that has ever existed contained similar percentages of stupid, lazy, and immoral people?

Part-Timer

The word 'hero' is used a lot nowadays.

So…

Are some people truly heroes? As in always?

Or do some people sometimes do heroic things…
…after which they get labeled as a hero for life?

Taxation with Representation

Suppose you were able to choose *exactly* what your tax dollars funded.

In this new system, a survey will arrive in the mail each year. You'll get to check boxes based on what projects you want your money to be used for.

Name two things you'd definitely want your taxes to fund.

And two things you'd never, ever use taxes to pay for.

Revenge Algebra

Complete the following equation.

Use no more than three words per blank.

_____

$+$

_____

$=$

A suitable punishment for a serial killer

Professionals

For each of the following, state whether or not you believe it's a low-class or shameful profession:

Garbage Removal

Pizza Delivery

Prostitution

Exotic Dancing

Professional Gambling

Weapons Manufacturer

Slum Landlord

Politician

Shoot for the Moon

Consider *everyone* else in the room with you. For each person, imagine a gift you'd like to give them. This gift can be anything: physical, emotional, spiritual, etc.

You can only give them one thing.

Your imagination is the only limit.

While They're Standing on a Cliff's Edge

Tomorrow you're going to find out that your spouse, fiancée, or lover has been cheating on you (vigorously) for many months.
If you had to choose, *how* would you want to learn about it?
Stumble across some emails?
Overhear a phone conversation?
Walk in on them while they're _____ing?
Some other way?

SPF 1,000,000

You deserve a break after answering piles of deep philosophical questions.

So...

If after your death you could live on as a vampire, would you?
You'd have to sip human blood and avoid the sun at all costs. But otherwise you'd be pretty much immortal.

AK-Forty Whatever

Imagine you can build a special kind of gun.
This gun has *unlimited* ammunition. You can load it with anything imaginable. You only get one choice.
Bullets. Lasers. Rockets.
Money. Medicine. Liquor.
Sweet dreams. Happiness. Orgasms.
Or something else.

What's in your gun?

Super Mega Awesome Eve

Invent a new national holiday.
No businesses will be open on this day.
Everyone will celebrate it.

What day of the year is your new holiday?
What's the day called?

Nutshells

Use one word in each blank:

_____

$+$

_____

$=$

The meaning of your life.

Love Fades

Consider your life's passion.

It's your goal, your end-game, the fire beneath your bottom.

Now...

Name something that could make you give this passion up.

Cyborgs

Imagine that technology exists to improve the human body via cybernetic upgrades.

In other words, for the right price, you can replace any portion of your body with a powerful and aesthetically pleasing robotic version.

Arms. Legs. Eyes. Organs. Et cetera.

Assume a 300% improvement for each upgrade.

Will you do it?

If so, which two body parts will you upgrade first?

You are Legend

Imagine the apocalypse has struck.

Every single human in the world has died.

Except you.

Aside from survive, what will you do during the long, empty years ahead of you?

Be specific.

Job Security

Suppose you could ingest a government-issued drug that would extend your life to approximately 200 years. You'll age only 25% as quickly if you take this drug.

The catch:
You'll have to work at the same job for the entirety of your life.
Would you take the drug?
What specific job or trade would you commit to mastering?

The Lifeboat

The old saying goes, "Women and children first." Meaning: if there's grave danger at hand, women and children should be rescued first.
The men be damned.

Do you agree with this premise?

The Island

In this exercise, you will build your perfect *sanctuary*.
You have only the following rules:
This place will be utterly removed from the rest of the universe
You will be the only inhabitant
It will be exactly one square mile in size
You can fill it with any landscapes, buildings, climate, and animals you desire, but no people
You will exist alone there for one-thousand years
Now build.

Worms in the Same Apple

Which of the following three people is the most vile?

Someone who manipulates *others* to do terrible things (Hitler, Stalin, Pol Pot.)

Someone who *willingly* does terrible things under the command or desire of another (Nazis, etc.)

Someone indifferent who stands by and does *nothing* while the innocent suffer?

Paranoia

Imagine someone in real life wants to kill you.
Who among your family, friends, frienemies, and
associates is this person most likely to be?
What's the most likely reason they'd want *you* gone?

Stereotypes

Do you believe a person's personality can be predicted
based on any of the following?

The music they listen to?
Their job?
The television shows they watch?
Their marital status?
Their looks?

The Artisan

Tomorrow you'll wake up with a new skill of your choice.
You'll be the best in the world at this one thing.
You can pick only one skill.

Choose this new talent carefully.

A Dark Day

For one single 24-hour day, you get to be an all-powerful *villain*.
During this day, the world is yours to rule.
You can act with total impunity.
You can satisfy any desire, no matter how depraved.
You can commit heinous crimes without being punished.
At the day's end, the world resets to normal and all the evil you *may or may not have done* vanishes from existence.
Describe your day in detail.
Think you'd enjoy it?

The Hunger

Every human being is born with the desire and the need to eat, drink, reproduce, and survive.
This instinct includes the will to compete for resources, to fight, and even to *destroy* if threatened.
Most of these instincts don't have to be taught.
They just are.
Knowing that, is there truly such a thing as *innocence*?

Woulda Shoulda Coulda

Should:
...politicians be allowed to receive money from special-interest groups?
...women be allowed to serve in military infantry divisions?
...most drugs be legalized?
...companies be allowed to fire employees for whatever reason they want?
...standardized testing be eliminated?
...fortune cookies be required to contain actual fortunes, not just random statements?

Have you ever suddenly and completely reversed your personal position regarding a major component of your life?
Such as: your religious beliefs, your political tendencies, or the way you felt about an important cultural/social event.
If so, explain in detail.
If not, why so hard-headed? ☺

Wash the Chalkboard

You've had a terrible accident.
All of your memories have been completely wiped out.
Except one.
Choose the one single memory you'd want to keep above all others.

The Slayer

Once.

That's all you get to use this next power.

Just one time, you get to slay anything in the world.

One person.

One thing.

One idea.

One way of life.

Anything you want: Dead

So, killer...

What'll it be?

Talk is Cheap

On a scale of $1.00 to $100.00, what value does an opinion have if it's stated by someone who *thinks* about the issue at hand, but *does* nothing?

Now estimate how many $1.00 opinions you've had in your life.

Seriously.

Tick or Tock

For each of the following proverbs or ideals, call out *yes* or *no*. Yes means you support it. No means you reject it or you're not so sure:

An eye for an eye

A picture is worth a thousand words

Expectation is the root of all heartache

History is written by men who have hanged heroes

Whatever we think, we become

After the game, kings and pawns go in the same box

Keep your friends close. Keep your enemies closer

Explain your answer to at least one of these.

Permanent Vacation

You're taking a trip.

A *loooooooooooong* trip.

It'll last for thirty years, during which you and *one* other person will travel together to and from an interstellar paradise.

No one else is coming with you.

So...

Who's your travel mate?

And why them?

Jesus F#^(ing Christ

Name a cause you'd be willing to become a martyr for.

As in *you die*, and the cause succeeds because of it.

A Price for Everything

If you could guarantee yourself a peaceful, healthy life (*not necessarily happy,* but peaceful and healthy) which of the following, *if any*, would you be willing to give up?

Sex
Alcohol
Your pets
Music
Good food
Family
Love

Open Heart Surgery

Tell everyone in the room about the *love of your life*.

It doesn't necessarily have to be a romantic love.

It can be a person, an animal, a passion…

If no one else is in the room with you, skip this question until a later time.

But if other people are here, open your heart for all to see.

Just be careful not to surprise your wife. ☺

The Exhumer

You've been given a one-time power.

With it, you can resurrect any one thing, as in bring it back to life.

It can be a person.

An animal.

An object.

An idea.

A tradition.

A cultural norm.

What will you resurrect?

Change your Underwear

Name at least one thing that makes you weak in the knees.

Explain why.

War Against Heaven

In John Milton's epic poem, Paradise Lost, the angel Lucifer rebels against his creator.

His motivations are jealousy (for his creator's love of mankind) as well as a desire to no longer spend *eternity* at his master's feet.

Given the situation, if *you* were in Lucifer's shoes (or hooves ☺) would you rebel?

Would you be jealous of your creator's love for an inferior species?

Would you wage war to be free of a master you believed didn't value you?

Rock, Paper, Machine Gun

Which one of the following three human traits is most powerful?

Intelligence (Our minds)
Emotions (Our passions)
Instincts (Our biology)

Answer for humanity at large and for yourself.

Bottom of the World

Imagine...
Your whole life, you've been looking for something. You've traveled far and wide. You've climbed mountains. You've plumbed the depths of the deepest ocean trench.
Finally one day you learn the object of your desire rests in the lowest chamber of the world's deepest cavern. You go there. You find a treasure chest. You unlock it. What's inside the chest?
What's the one thing you've searched your whole life for, but you've never been able to find?

Told You So

If you had to guess, which of the following events is most likely to cause the end of the human race?

A meteor strikes Earth

A terrible disease infects everyone

Nuclear warfare

We invent artificial intelligence, which in turn destroys us

The Sun goes supernova

A higher being wipes us out

Aliens

Or _____

Democrachy

Suppose you are the ruler of a new nation.

In general terms, what kind of government will you institute?

(Create a new government type if you want.)

Sensory Deprivation

What would you miss most if tomorrow you:

Went suddenly blind

Lost your sense of smell

Went deaf

Lost your ability to taste

Lost all sensation of touch

?

Assume each condition is permanent.
Be specific about what you'd miss.

One _____ to Rule them All

Is it morally wrong to desire power?

The Definition of Darkness

It's a fair assumption that almost all people have had dark thoughts from time to time.
Racist thoughts.
Misogynistic thoughts.
Dark sexual fantasies.
Perhaps even visions of destruction, murder, and violence.
So...
What's your personal darkness?
What's something you've thought of, accidentally or otherwise, that might terrify the people who know you?

101 Questions for the End of the World

In a cage match between science and philosophy, who wins?

Everyone.

Butterflies

The *Butterfly Effect* is defined as:

The phenomenon whereby a small change within a complex system can have large effects elsewhere.

In other words, a butterfly making the smallest alteration in the wind might set off a chain reaction causing a hurricane two weeks later on the opposite side of the world.

Or something like that.

Think of an important event that has taken place during your life or the life of someone you know.

Now think about how that event came to be.

What was the butterfly?

Does Your Head Hurt Yet?

Suppose that when you wake up tomorrow, you know everything.

As in *everything*.

Everything that has ever happened.

Everything that is currently happening.

And everything that will happen.

And you'll know the reasons *why*.

In theory, this would enable you to answer any possible question the universe could pose to you.

Or would it?

If you knew everything there was to know, is it possible such knowledge would only create *more* questions?

Meaning you could never truly know everything, because for each answer you had, a new question would pop into existence... infinitely.

Well?

Chills

Cryonics.

In other words, freezing terminally ill humans with the aim of restoring them in the future using medical advances that have yet to be invented.

For purposes of this question, let's assume someday medicine will reach a point of actually being able to thaw and restore a cryonically frozen person.

Now imagine you're terminally ill. Would you allow yourself to be cryonically frozen if the restoration process were guaranteed to be invented in:

50 years?

150 Years?

1,000 years?

10,000 years?

The Pilgrimage

On a long enough timeline (10,000 years or so) it's *possible* humanity might develop and utilize the technology to spread our species to other planets.

Meaning:

Interstellar travel

Terraforming

Planetary colonization

And given an even longer timeline (10,000,000 years or so) it's possible (however unlikely) humanity might be able to occupy every habitable planet in our galaxy, maybe even beyond.

Suppose it were *definitely* possible.
At what point do you believe humanity would be wise to stop its own spread?
Any chance you'd say *never*?

No Accidents?

Scientifically speaking, everything happens for a reason.
Literally.
Meaning on a physical level every interaction in the universe has a definitive physical cause, no matter how obscure.
And thus, universally speaking, there are no accidents.
No luck.
No true randomization.
But…
Things like human emotions, impulses, and ideas don't *necessarily* fit into any standard scientific construct.
Meaning the reasons *behind* several of humanity's physical actions aren't exactly known.
Meaning human activity can disrupt the universe's interactions.
So are there accidents after all?

The Doctrine of Double Effect

There's a runaway train.

Aboard it are five people.

Thing is: you're standing at a lever which, if you pull it, will switch the tracks and save the five people.

But...

If you pull the lever, you'll redirect the train to a portion of the tracks where a woman and her baby are standing, totally unaware of what's about to happen.

If you *don't* pull the lever, five people will die through your inaction.

If you *do* pull the lever, two people will die due to your direct action.

Which choice is correct?

Soul Food

First, a few definitions:

Soul - sōl/ (noun) - the spiritual or immaterial part of a human being or animal, regarded as *immortal*.

Mind - mīnd/ the component of a human being or animal that enables them to be aware of the world and their experiences.

Despite our many technological advances, one thing science has yet to define and/or locate proof for is the existence of souls.

Also hard to define are the processes of thoughts, feelings, and perceptions.

So which is more likely?

Souls and consciousness are physics-based, and will one day be completely explainable by science?

Or

Souls and consciousness operate under separate parameters and are not bound by any known laws of physics?

The Alpha and the Omega

Is it possible:

The only true purpose for all life everywhere is to survive and replicate?

The Fear of Consequence

Morality is a slippery concept.

Good and evil have shaky foundations.

Point is: what's considered good and noble in one culture isn't always viewed likewise in another culture. Moreover, left without supervision, individuals tend to take a lot more liberties with morality. If the authorities aren't around, people will assault, loot, and murder more than if there's a police car nearby with its lights blazing.

Which begs the question:

Are people 'moral' only because they fear punishment if they're not?

If the concepts of authority and law didn't exist, and no punishment awaited disturbers of the peace, would the world gravitate toward violence and entropy?

If so, does that mean morality only exists beneath the fear of consequence?

The Paradox

Socrates, famed Greek philosopher, once posed that any man who believed himself wise was in fact *not* wise at all.

His reasoning was that any man who boasted of his own wisdom had no true awareness of his ignorance, and therefore could not be considered wise.

Meaning, paradoxically, Socrates might have been among the wisest men of his time, if only because he understood and accepted the vastness of his own ignorance.

The question:

Does being wise mean acknowledging one's own ignorance?

Taking up Space

The adage goes something like:
"We're a part of something larger than ourselves."
The implication is that humanity is a tiny, yet important cog in the universal machine.
But what if we're not?

Think about this: nearly *all* the physical places in the universe are inhospitable, if not downright hostile to living organisms. Not just to human life, but to all life. And not just a little hostile, but completely, utterly deadly.
The implication is that the universe might not appreciate life as much as *life* appreciates life.
So...
Are we truly a part of something larger than ourselves?
Or are we just taking up space?

Dictionary.You

Define the difference between *science* and *religion*.

The Wrath of _____?

In many religions, there exists a day of reckoning. A day of extreme destruction, cleansing, or a spontaneous revelation of a new age of consciousness.

The theological term for the study of these end-time events is known as *eschatology*.

But eschatology is a little tricky to pronounce, so for now we'll call it Judgment Day.

Now...

Given that in most religions, an imperfect humanity was created by a perfect divine being, is it *fair* for humanity to be judged, destroyed, or otherwise altered on a worldwide scale?

Wouldn't the more logical action be to judge the *creator* of our imperfect system, rather than its temporary occupants?

Worth

Considering the universe and all the people, places, and things within it, assign a tangible monetary value to the following people, and do so as objectively as you can.

Your child

An adult stranger on the street

Someone else's newborn baby

A murderer

Yourself

The value must be exact. It can be any value between $0.00 and $100,000,000,000,000 (that's one-hundred trillion.)

Worth More

Now let's up the ante.

For purposes of this exercise, we're making up a point scale.

We'll call it the *Universal Point System*.

The system goes from 0-10. A *zero* means no value at all to the universe. A *ten* means the highest possible universal value. A *five* is somewhere in the middle.

For each of the following, assign a value based on the Universal Point System:

A drop of water

The planet Mars

A single human being

A blue whale

Earth

A star

A galaxy

Gravity

Game of Thrones

Choose which king you would prefer to rule the kingdom in which you live:

The Sword: *Strong, decisive, militarily powerful, yet prone to rash judgment*

The Peacemaker: *Calm, benevolent, generous, yet unwilling to take risks*

The Lawyer: *Highly intelligent, even-handed, able to solve complex problems, yet entirely non-emotional*

The Philosopher: *Visionary, wise, deeply thoughtful, yet not at all concerned with tradition or religion*

And which king would you *least* want on the throne?

The Grand Illusion

Ultimately, we have no objective proof that anything we know or experience is real.

However improbable, it's possible that due to our limited scope of sensory input, we exist in an illusion or machine controlled by an entity who exists beyond our perception.

If this were true, would you want to know?

Really?

It's All Gravity's Fault

Modern science does a spectacular job of explaining the exquisitely precise laws of physics governing our universe.

But...

There's no real way for anyone to know why our universe is governed by *these* laws as opposed to, say, a *completely different set* of physics laws. It's an unanswerable question.

In your mind, does this:

Indicate the intelligent design of our universe?
Mean we've still got a lot to learn scientifically?
Indicate a certain amount of randomness in terms of the universe's creation?
It's pointless to speculate on why we follow one set of physics laws instead of another?

(You can answer with more than one of these.)

The Value of a Thing

In many instances, science is an expensive habit.
Educating, experimenting, and exploring consumes resources and time.
In particular, space exploration costs *lots* of resources and time.
Some of these ventures yield amazing technological advances.
Others reveal key information about the universe we live in.
And still others try, but fail to accomplish much.
In your personal view, is...
The primary purpose of science to improve the quality of our lives and increase our level of convenience?
OR
The primary purpose of science to educate and enlighten us about the nature of our existence?

Are there any specific fields of science you would eliminate due to high cost and low yield?

Strung Together

In a nutshell, String Theory (aka: the Theory of Everything) goes something like:
All of the different *fundamental* particles of the universe are really just different manifestations of one basic object: *a string*. Not literally a piece of string, but a tiny, tiny object that moves similarly to one.

Meaning that if the string moves in one way, it's a *proton*. But if it moves in a different way, it's an *electron*.

In your mind, if this theory turns out to be true, does this mean the entire universe and everything in it is made of the same exact stuff?

Quotable

Which of the following do you believe?

"Words exist because of meaning. Once you've found the meaning, you can forget the words." ~ Chuang Tzu

"Man is the only animal who enjoys the consolation of believing in a next life. All other animals enjoy the consolation of not worrying about it." ~ Robert Brault

"The tighter you squeeze, the less you have."

"The opposite of a correct statement is a false statement. But the opposite of a profound truth may well be another profound truth." ~ Niels Bohr

"You have to do it by yourself. And you can't do it alone."

~ Martin Rutte

There is no *Slightly*

Chaos Theory goes a little something like this:
In complex systems (a machine, the Earth, or even the entire universe) having *slightly* different initial conditions will yield dramatically different outcomes.
For example; if you cloned Earth down to the last atom and started their existences at the exact same time, but made the temperature on Earth #2 just 0.01% warmer, over the course of time the two planets would end looking very different.

Now...

Suppose you didn't mess around and made the two Earths *exactly* the same, including the starting temperature, would they look the same 1,000 years later?

Obviously the answer is unknowable.

But what do you *think*?

Nothing is Not

Nothing is less than what most people think it is.

Nothing isn't an abyss. An abyss is *something*.

Nothing isn't a vast empty space. A vast empty space is *something*.

Nothing is *nothing*. No space. No emptiness. And most importantly, no potential to ever be anything more than nothing.

Now comes the question of belief versus truth.

Why is there *something* instead of nothing? Why does our inconceivably huge universe exist instead of nothing? Why are there stars, planets, human beings, paper clips, and Pokemon…instead of nothing?

If your answer is anything other than, "I don't know," you have some serious explaining to do.

Deterministic Dilemma

Fact: atoms behave in a probabilistic manner (meaning they *usually* do *exactly* what they're predicted to do.)
Fact: our brains are made up entirely of atoms
Fact: recent scientific advances using Functional Magnetic Resonance Imaging (fMRI's) have determined that our brains typically make decisions several seconds *before* we're even aware of it

So...

If our brains, being made of things that do exactly what they're supposed to do, and being capable of deciding things before their hosts (us) are aware of it...
...does free will exist?

No Time for an Investigation

Aside from discussing the mundane, is it arrogance when someone says they *know* the truth beyond a shadow of a doubt?

I'm Just Here for the Free Drinks

When someone asks, *"What is the meaning of life?"* perhaps they should be a bit more specific.

Because…really…there are multiple conditions to such a question.

Are they asking, *"What is the meaning of ALL our lives?"* which indicates a grand, unified purpose behind humanity's existence.

Or maybe they're trying to ask, *"What is the meaning of YOUR life?"* which implies our purposes are self-created.

Or, given that our individual views on the subject are extremely subjective, maybe all they're really asking is, *"Is there any meaning to ANYONE's life?"*

So the question is: which one of these questions is the right one to ask?

Is the meaning of life self-generated, as in it's different for each individual living thing?

or

Do all lives have similar meaning?

Collisions

Bill and Sara have just had an argument.

The day after the argument, Sara calls Bill to tell him he really hurt her feelings. She says she's upset, and will be for a while.

Bill rejects this notion. He claims steadfastly that he didn't hurt Sara's feelings, and states she has no reason to be upset.

This is clearly a collision of perception.

Both Sara and Bill firmly believe their positions.

Objectively speaking, did Bill actually hurt Sara's feelings?

Explain your answer.

I'm Done with this Test

Nirvana

It literally means: *blown out (like a candle) or 'life extinguished.'*

It's not meant as a negative thing. It's meant to represent a profound peace of mind, an utter release from egotism, and an end to the cycle of karma.

In technical terms, Nirvana is specific to the Hindu faith. It's a state of consciousness ending the cycle of reincarnation.

But let's move beyond that.

Let's suppose Nirvana is a worthwhile state of mind to reach for *anyone* with *any* belief system. Utter peace of mind sounds pretty ideal, right?

Imagine Nirvana is a legitimate thing, and that you have but one single lifetime (your life, right now) in which to reach it.

Would you try to achieve it?

And how would you go about doing so?

You are (*only*) what you Eat

Materialism is a doctrine stating that all things in the universe, without exception, are bound by the laws of physics.
In other words, there is matter, and there are forces that make matter do stuff, but there are no meta-forces guiding the matter. Also, materialism disallows separation of consciousness from all other physical interactions.
Meaning our brains and our thoughts are *one*.
Now, the argument starter:
If materialism is valid, does it…
A. *Imply there can be no intelligent creation?*
B. *Mean free will does not exist?*
C. *Imply a certain pointlessness to our lives?*
D. *Eliminate the need for morality?*
E. *Not bother you in the slightest?*
(Choose as many as you like.)

Manifest Reality

Conversely to materialism, we have *idealism*. Idealism asserts that our entire existence is a mental, consciousness-driven one, and therefore is utterly mutable by our perception of reality.

While many branches of idealism recognize the existence of physical matter, nearly all of them declare that our reality is *conceived* by our consciousness.

Now, the argument continues:

If idealism is valid, does it...

A. *Mean Santa Claus exists if a child truly believes in him?*

B. *Imply the likelihood of intelligent design?*

C. *Allow each individual his or her own self-constructed reality?*

D. *Allow for events beyond those explained by physics, including supernatural and spiritual occurrences?*

E. *Mean individuals' separate realities are constantly crashing into each other?*

Why Bother?

There exist many humans in the world who maintain a firm belief in determinism.

In fate.

In destiny.

In other words, they feel that no matter what actions humanity takes, the outcome of our existence will be the same.

Why then would anyone who believes this philosophy:

Look when crossing the road?

Go to work?

Eat?

Ever do anything?

Or were they pre-determined to do or not do these things?

A Party for the Gods?

If in fact there was a Big Bang, meaning the almost impossibly powerful explosion that birthed our universe...

...what happened before it?

Or maybe there was no *before*?

Born Yesterday

It's known that extremely high velocities slow time relative to objects not moving as fast.

Meaning that for an object traveling at high speed through space, time will move slower than for an object sitting on Earth.

With this in mind, is it possible that the light from our universe's earliest stars (*which travels at 671,000,000 mph*) was only just created *moments ago*?

Think about it.

Changeling

Every moment of your waking life, you are *experiencing*.

Witnessing.

Learning.

Forgetting.

Even if you're not aware of it, it's happening to you right now. Therefore, scientifically speaking, your brain isn't in the same physical state it was at the beginning of this sentence.

So…

If your state of mind is ever-changing, and if your exact experience of life is always different than it was just moments ago…

Are you a different person now than you were thirty seconds ago?

What Will Be, Will Be

Imagine a model of the universe in which there is *definitely* a god.

For many, this won't require any imagining at all.

Now…

In most religions, the god or gods are not only all-powerful, but also all-knowing.

Meaning they know what *has* happened, what *is* happening, and what *will* happen.

And scientifically speaking, if the future is a known absolute, then by its very nature it cannot be changed.

Meaning everything humanity does is fixed.

Fated.

Predetermined.

Therefore, if in this model all actions are predetermined, why should any human be held responsible for any action ever?

Including the horrible ones…

We Can But Try

Objective - /əbˈjektiv/ -

A person who is not influenced by personal feelings or opinions when considering and representing facts.

In other words, completely unbiased and unmoved by anything but the truth.

Can a person ever be 100% objective?

If so, can they do it *always* for *everything*?

Inconceivable

The universe

Life

Gravity

Consciousness

These all have something in common:
They're all parts of our reality, but also have an unknowable source of creation.
In other words, we don't know *why* these things exist.
Lacking complete data on the origin of these (and many more things) does humanity have little choice other than to declare these phenomena are *real* and move on?

Or is it our scientific duty to pursue a reason why these things exist?

Play the Percentages

How much of your total knowledge is *firsthand*? As in, *you* saw it. *You* experimented with it. *You* understand it due to *your* direct contact with it.

Examples:

You threw a ball into the air and watched it fall down.

Thus...gravity

You hit your thumb with a hammer.

Thus...force and pain

You observed a tree react to the climate for an entire year.

Thus...seasons

And how much of your knowledge is either derived from inference or hearsay? As in, someone else told you about an event, you saw the event on TV, or you read about it in a magazine?

Give percentages for your direct and indirect knowledge.

Bonus question: is a piece of knowledge gained indirectly nothing more than a *belief* until you witness it or prove it firsthand?

Turing

The Turing Test - designed by Alan Turing, an assessment of a computer or artificial intelligence performed by a human with the intent of gauging whether or not the computer or AI possesses consciousness.
Got that?
Now let's go beyond computers.
Suppose you performed this test *not* on an attempted AI or a computer, but on *another human*.
Now, two crazy questions:
Is there anyone you've met in your life who would fail this test, thus failing to convince you they had consciousness?
And...
Can consciousness be reproduced? And if not, does that prove the existence of souls?

The Conjecture Clock

First, here's a few interesting measurements of time:

Attosecond – Currently the smallest division of time. Approx 10^{-18} seconds.

Megasecond – Approx 11.6 days

Galactic Year – The time it takes for the Sun to orbit once around the Milky Way's center. Approx 230 million years.

Exasecond – Approx 31.7×10^9 years. (more than twice the age of the universe.)

Now, the real question:

Does time exist?

Or is it simply a human construct?

When answering, *take your time.*

Thought Police

Is it immoral to be willfully ignorant?

Two Worlds

Humans experience some pretty strange phenomena.
Things like *déjà vu, synchronicity, placebo effects.*
Despite the belief (or hope) that these effects are spiritual or otherwise outside the realm of explanation, most of them have causes rooted in science.
And yet...
A few phenomena exist that have yet to be fully explained.
Things like *ghosts, past-life memory, ESP.*
Which leaves us with three distinct possibilities:
These things don't really exist. People make them up.
These things do exist, but have scientific reasons we've yet to find.
These things do exist, but have causes outside the realm of science.
Which one do you think is most likely?
And why?

Plato's Play-Doh

Plato, famed philosopher, once suggested that a bad democracy is *better* than a bad tyrant.

Mostly because in a bad democracy *everyone* is participating in the badness.

Think hard.

Would you prefer a lone, bad tyrant or a bad, malfunctioning democracy?

Or are both equally horrible?

The Sun will Rise Tomorrow. Won't It?

If you can, name three things or phenomena it's acceptable to *believe* in without having actual objective proof of that thing or phenomenon's existence.

It's *Your* Turn

"Where must we go, we who wander this wasteland, in search of our better selves."

No, it's not a quote from Socrates, Plato, or Neil deGrasse Tyson.
It's from George Miller, Hollywood director.
But it's still a good question.
Specifically, it's similar to the author of this book's big life question:
Where must we go and what must we do to find our better selves?
Now it's your turn.
The universe has agreed to answer one single question of your choosing.

What will your question be?

The Wellspring

Must all knowledge be derived from using our actual senses to learn it?

Meaning, must we be able to see, hear, touch, taste, or smell something in order to know *without doubt* something is real and true?

Examples:

We touch a baseball and count its seams - therefore the ball is real

We feel a cold wind blowing over us - therefore the wind is real

We see other people walking by - therefore they are real

Or...

Can *some* knowledge, such as the existence of a god or the presence of a ghost, be assumed without sensory proof?

Explain.

Need to Know Basis

Buddhism

It's a philosophy in which the primary purpose of life is to earn freedom from restlessness, karma, and unease. One method some Buddhists seek peace is to avoid questions that focus on non-physical matters. They believe a focus on the *metaphysical* aspects of life only leads to deeper speculation, thus more questions, thus greater *unease*.

Unease which is directly oppositional to their peace-of-mind goal.

Would most of humanity find happiness easier to achieve if they abandoned the metaphysical questions they have about life and the universe?

Yesterday versus Tomorrow

Regret

Worry

Two powerful states of mind that have long influenced humanity.

Regret arises from past events, which, barring the invention of time travel, are unchangeable.

Worry stems from fear of what the future may bring. The future may or may not be changeable, depending on your view of determinism, but it doesn't appear to stop people from worrying.

So...

Which one is a greater influence on humanity?

Or are they equal?

Apollo and Dionysus

In Greek mythology:

Apollo represents reason and rationality.

While *Dionysus* is the god of chaos and irrationality.

This dichotomy may have originated in myth, and yet like most Greek stories, it has its roots in human philosophy.

The point:

Humans are capable of great reason and rational thought, and yet are just as often inspired to deeds of great passion, some constructive and some perhaps not so constructive.

The questions:

Does this duality accurately describe humanity?

Are we split into two powerful spheres, reason and emotion?

Or are these things not truly separate, and actually just different expressions generated from a single source: our brains?

The Duplicant

Scientists have begun the study of mapping the human brain and 'creating' memories via experimentation using the portion of our brain known as the hippocampus.

Also...

Studies have been conducted to test the viability of *storing* human memories indefinitely.

Using such methods, it's possible one day we'll be able to erase the memories we don't want, store the ones we need for later, and perhaps even create memories of things that never happened.

Now suppose you were in a car accident.

You were rendered comatose for five years.

You were declared *brain-dead*.

But...before the accident, you had all your memories stored in a computer. And upon waking in a stupor from your coma, these memories were successfully re-installed into your brain.

Is the person waking up from the coma *you*? Or *not* you?

Promethean

Does humanity possess a mandate to continue our technological development until its utter end?

Meaning, should we advance ourselves technologically as far and as fast as possible?

If not, at what point during our advancement should we consider slowing down or stopping?

No Place Like Home

Reincarnation

It's a prime tenet of the Hindu religion, and is referred to peripherally by several other faiths.

The idea:

Every human and animal has a soul.

Upon death of the body, the soul can inhabit a new body, thus beginning life anew.

Let's imagine for a moment reincarnation is absolute fact.

Now suppose, due to some universal cataclysm, all life were permanently extinguished. It's entirely possible, even if it happens many, many years from today.

If all the souls in existence were suddenly rendered homeless (no bodies to inhabit) would they simply wander the universe for eternity?

And would that make the entire reincarnation concept pointless?

In the words of Pyrrho

Is it possible humanity can never really know the truth of *anything*?

As in, due to our limited scope of sensory input and highly human-centric perspective, can we only ever know the *appearance* of things, and never the reality of them?

It is NOT what it is.

"It is what it is."

You've probably heard this declaration a lot. And if you haven't, you probably will.

But scientifically, this statement is probably always wrong.

Earlier in this book, we discussed how the human mind is never in the same physical state as it was moments ago.

Now let's apply this to the entire universe.

Given that every piece of matter, from tiny atomic particles all the way up to stars and galaxies, is actually moving, shifting, and oscillating at all times, nothing we perceive is ever the same twice.

Electrons, cells, rocks, trees, rivers, asteroids, people...

Always moving at all times, even if imperceptible to the naked eye.

Does this mean you and everything you know is never the same as it was a millisecond ago?

Is everything *not* what it is?

Deus Ex

Imagine at some point in the future humanity successfully creates AI.
Meaning we design an artificial intelligence that demonstrates a clear ability for consciousness.

If we create such a thing, are we gods?

Not Quite What You Were Hoping

There exist numerous theories regarding the meaning of life.

Some predict a divine afterlife.

Others believe in infinite recycling of our souls.

Some believe in very specific versions of heaven and hell.

And still others say there's no meaning at all.

Everyone is guessing.

No one really knows.

Even so, the most common perception is that *if* there is a meaning, it's probably a positive or at worst a neutral one.

But…

What if humanity one day learned our purpose is nefarious?

That perhaps humanity (or even all life) was engineered for a negative purpose?

Is it possible?

If you learned such a thing were the truth, what would you do?

Finally, a Simple Question

Given the choice, would you rather know *how* the universe works, meaning you'd understand all the hard science behind each and every interaction taking place in our existence?

or

Would you prefer to know *why* our universe and all the individual objects within it exist, meaning you'd grasp the *purpose* behind everything?

Explain your reasons.

Something in our DNA

Are humans nothing more than highly intelligent animals?

Or are we something special?

Something unique?

Are we a species deserving a larger share of the world, maybe even the universe?

Climbing Up the Chain

The science behind humanity's physical evolution is well-known.

We've unlocked the human genome.

We've studied macro and micro evolution of our species.

Generally speaking, we've come to understand how and why our skeletons, organs, and brains are the way they are.

And then there's *conscious evolution*.

It's a theory asserting that due to our technology and understanding of the human condition, we now have the ability to control our future evolution.

Meaning we can either cooperate to improve our bodies, minds, and societies...

...or compete and ultimately destroy ourselves.

Imagine tomorrow the entire world agreed to *cooperate*.

What should be our first goal in the next step of human evolution?

Cartesian

"If you would be a real seeker after truth, it is necessary that at least once in your life you doubt, as far as possible, all things." ~ Rene Descartes

It's a pretty powerful statement.
Do you agree with it?
Why or why not?

Odds are: We're *all* Wrong

At the time of this book being published, the human population of Earth exceeds seven billion.

And among these seven billion can be found *thousands* of varying viewpoints, including:

Different religions

Different views of morality

Different social and political beliefs

Scientifically speaking, if *one* particular combination of beliefs happens to be absolutely true, then by rule the other belief systems *cannot* be true.

Meaning a huge percentage of the world's population holds a belief system that is likely incompatible with reality.

So...

Are you among the small percentage of people whose beliefs are true?

If so, does this mean everyone who disagrees with you is wrong?

That's *MY* Throne

Typically when discussing humanity's effect on Earth and all the lives and ecosystems on it, we tend to talk in the negative.

We use words like:

Wasteful

Dirty

Greedy

Destructive

Moreover, we tend to emphasize that humans have these qualities, and that other animals (often our victims) do not.

And yet, one must wonder…

If other species had the same tools we humans have (large brains, opposable thumbs, language, et cetera) is it likely they would take our place as Earth's dominant life form…

…and be just as destructive as we are?

The Monk

Is it possible for a human being to live without *believing* anything?

Meaning this person would have:

No opinions

No claims to knowledge without hard physical evidence

No spirituality

No philosophy beyond the material world

No religion

If someone could pull this off, would it be admirable?

Or is this an impossible state of mind for a human to achieve?

The Human Lens

Sixth senses aside, *everything* you know about the world, you know through the subjective lens of your human brain.

Meaning you only truly know what you see, hear, smell, touch, and taste.

You'll never know what it's like to see the world in the same way a cat does, or a bird, or a whale, or a bacterium.

Meaning you'll only ever experience the universe from a human point of view.

And more specifically, *your* human point of view.

So...

Does this mean your experience of reality is unique, almost isolated in its *filtered-through-a-human-lens* nature?

Or does this mean that physical reality itself is different for every single living thing?

If it Hasn't Happened by Now...

If, in the entire history of humanity, no one has yet been visited by someone from the future, does that mean time travel won't ever be invented?

Cardinals

The old saying goes:
"Do unto others as you would have them do unto you."

It sounds great, right?

Until you realize that many people wouldn't mind having some *pretty strange and awful* things done to themselves. Which, in theory, gives them license to do those same things to someone else.
Now let's break this cardinal rule.
If you can, name at least one thing you'd like done to yourself that would invalidate the *'do unto others'* adage.

And is there any such thing as a perfect ethical system?

I *Am.*

Human self-identity is largely a function of perception. Meaning how we see ourselves appears to dictate who we are.

Or at least, who we *think* we are.

Moreover, if someone perceives themselves as being something (smart, stupid, strong, weak, et cetera) the tendency is for them to gravitate toward becoming that thing.

In their own eyes. And possibly in the eyes of others.

Now suppose you took two people and switched their minds into each other's bodies.

Meaning Jim's mind goes into Sara's body, and vise versa.

A year goes by after the switch.

Jim and Sara are now accustomed to being in their new bodies.

But are they still Jim and Sara?

Or, due to having completely different perceptions based on their new bodies, are they different people entirely?

If Time Were a River

We've already discussed time travel.
And the likelihood that it's impossible.
But...
Suppose it were possible.

Imagine you had the capability to enter a time machine and go back one-hundred and fifty years, roughly to the same time period during which your great-great-great-great grandfather lived.

If you were to kill your great-great-great-great grandfather, you would in effect prevent all his offspring from existing, including yourself.

So would that mean you couldn't have time traveled in the first place?

Leaving

Just a few of the potentially lethal obstacles preventing easy space travel are:

Time

Cosmic radiation (small, dangerous particles traveling at high speed)

Space debris (traveling at high speed)

Life-support

Psychological effects of prolonged time in deep space

Exiting and entering planetary atmospheres

Given these, and given the vast amount of resources needed to even *attempt* to conquer these, answer this one question:

Is deep space travel a worthwhile human endeavor?

Just Because

Set your existing religious beliefs aside for a moment.

Imagine, no matter what you normally believe, there is definitely *one singular divine being* responsible for the creation of all things.

A god, a goddess, or whatever.

Now...

Justify why this divine being created:

Viruses

Cancer

Black holes

Serial killers

Good luck.

Prior to recent scientific research, it was believed that every cell in the human body replaced itself approximately every 7-10 years.

This is now known to be false.

The neurons in the human brain are not replaced during our entire lifespan.

Even so...

Most of our other tissues are replaced, including fat, muscle, and bone cells.

Meaning the vast majority of your body is not the same body you had 10 years ago.

Does this mean you are a different person than you were a decade ago?

Or does this mean humans are in fact the contents of their minds, and not their bodies?

The Ache

It has been said that most of humanity suffers from a
permanent type of pain.
Not a physical pain, mind you, but a subtle unease and
dissatisfaction due to our mortality, our unmet desires,
and the transient nature of all life's experiences. (i.e., all
good things must come to an end.)

Do you feel it?
The quiet ache beneath your heart?
However small or large it might be?

Is it possible to cure?

Izms

From the following, choose which one(s) you associate with your personal philosophy of life:

Cynicism - The purpose of life is to live with virtue and in harmony with nature (not what you thought it meant, is it? ☺)

Agnosticism – Humanity knows nothing beyond that which it can touch

Pragmatism – The most valuable things are tangible and practical

Hedonism – Life's purpose is to pursue pleasure

Capitalism – Life's purpose is accumulate wealth for the benefit of yourself and your family

Theocentrism – God is a central fact of our existence

Nihilism - Life is without objective meaning, purpose, or value

Existentialism – The universe is unknowable, yet humans still have individual purpose and responsibility

A Shallow Grave

Famed scientist Stephen Hawking twice pronounced,
"Philosophy is dead."
His assertion is that the pursuits of philosophy, in particular physics, have not maintained pace with modern science.

Given the rapid advance of science, technology, and our ever increasing understanding of the universe's makeup, is traditional philosophy still a worthwhile pursuit?

Or should we focus all our questions on hard science?

Holy ____!

Objectively speaking, it's logical to say that the most honest answer to many of life's deep questions is: "I don't know."

Is there life after death? – I don't know.
Are there aliens out there? – I don't know.
Is there a god? – I don't know.

In each case, *I don't know* is a responsible reply. And by this reasoning, both atheists and devout believers appear presumptive in their beliefs about the workings and creation of our universe.
Perhaps only the agnostic point of view is truly honest. That said, for each of the above three questions, what do you think is the *most likely* answer? And why?

Disaster Porn

Y2K

The end of the Mayan Calendar

Nostradamus

Polar ice caps melting

The theme here is obvious. For hundreds of years, scholars, theologists, and even scientists have attempted to predict the end of the world.

None have been correct. Yet.

Regardless, many humans appear concerned with how, when, and why Doomsday will arrive.

Perhaps more important than trying to guess when the world will end is understanding the psychology behind humanity's morbid obsession with it.

The question: does human interest in the idea of Doomsday reflect a certain unhappiness with the world?

I ♡ Sagittarians

Astronomy – the science of celestial objects, space, and the physical universe

Astrology - the study of the movements and relative positions of celestial bodies interpreted as having an influence on human affairs and the natural world

So...

You probably know:

...the stars contained in constellations are often millions of light years apart, having no real relation to one another

...planets and other celestial bodies have no known personality traits

Even so, several modern societies practice and follow *astrology*, identifying themselves in relation to constellations, attributing moods and life events to the movements of planetary bodies.

Why do you think this is?

The Vat

Imagine you and I and every other person in the world aren't really alive.
We're just constructs in a system created by beings (or machines) who exist beyond our collective consciousness.
They're watching us. Controlling what we experience.
Setting the parameters for all the physics we hold dear.
Don't laugh.
It's entirely possible, even if improbable.
The question is:
If tomorrow you became aware of this fact, meaning you saw a glitch in the system and became aware of humanity's condition, would it matter?
To you personally?
To anyone?

Lego Babies

Within the next hundred years (likely much sooner) it's probable that sufficient technology will exist for *all* parents to decide *every* aspect of their child's genetic disposition.

This means parents will be able to decide and/or edit their child's:

Sex

Skin tone

Height

Weight

And much more.

Suppose you're having a kid and this technology is readily available.

Do you use it? And to what extent?

Right or Might

Is it better to:

Do the right thing and fail?

or

Do the wrong thing and succeed?

(Note: to answer this question, you'll have to assume that such things as right and wrong actually exist.)

Think Twice

French author André Gide once posed:

"Believe those who are seeking the truth; doubt those who find it."

Do you agree with this statement?

Have you ever believed you knew a significant truth, only to find out later you were wrong?

And are you more objective as a result of that experience?

Context is Everything

Thomas Jefferson's famous quote goes likewise:

"...that all men are created equal, that they are endowed by their creator with certain unalienable rights, that among these are life, liberty and the pursuit of happiness."

But are all humans truly created *equal*?

It's probable Jefferson didn't mean to imply that everyone is equal physically, but rather that the inherent value of a human being is the same no matter their origin, belief system, or life circumstance.

But again, are *all* humans equal?

Is a cruel, vicious, murdering war-criminal as valuable as a kindly philanthropist?

Is a brain-dead person, though still human in physical terms, as valuable as a healthy, adorable infant?

Or is the value of a single human life completely subjective, and utterly different depending on perspective, therefore negating the concept of equality?

Magnetic Personalities

Describe whether or not you agree with the following two sentences:

Some people are drawn closer to one another by a nameless magnetism.

Others are divided by an unreasonable dislike.

It's not very scientific, given that humans aren't known to be particularly magnetic.
So if you believe both sentences are true, explain why.

Sartre's Sabre

Given that the answer to most of life's questions (and indeed most of the questions in this book) are, "*I don't know*,"

...is it fair to say:

Since we have no evidence of the correct answers to life's great mysteries, are we free to create our own?

Ascetic

Wine

Liquor

Sex

Fine foods

Drugs

Just some of the things humanity indulges in.

For the sake of this question, let us assume none of these things are *bad* or *immoral* by themselves. While it's possible the people overindulging in them might do harmful acts, the actual wine, sex, food, et cetera aren't to blame.

That said, is a person who indulges in *none* of these a stronger person morally than someone who indulges in them often?

Does denial (or severe limiting) of one's indulgences make a person better?

Or do indulgences have no bearing on a person's *goodness*?

Embracing the Shadow

Suppose the following three statements were true beyond any shadow of a doubt:

No gods or goddesses exist

After death, there is only nothingness

Humanity is inconsequential to the universe

Given these statements, what incentive would a human have to be good instead of evil?

And…

Would good and evil even be relevant?

Sub-Objective

Are you what you *are*?
Meaning, are you flesh, bones, and synapses firing? An animal through and through?

Or are you what you *think* you are?
Meaning, are you the sum of your experiences, your instincts, and your thoughts? A temporary inhabitant of a borrowed body?

Or are you both?

Like Father, Like Son

In most religions, gods and goddesses are depicted as having human or animal characteristics.
Many times these physical traits are exaggerated. But most often they resemble *something* that appears on Earth (people, elephants, serpents, et cetera.)
For the sake of this question, *no matter your religion*, suppose gods and goddesses are absolutely real.

Do you believe a god presents itself as a familiar earthlike thing in order to make you (their worshipper) feel more comfortable?

Or is it more likely, assuming gods exist, that all the statues and pictures made by humans probably don't resemble what actual divine beings look like at all?

End of Ages

Is it possible (or even probable) that ages from now, much of the science and philosophy we now take for truth will be revealed as false, and a newer, truer system of knowledge be put into place?
In other words, could it be a lot of the things we *think* we know are completely wrong?
Also…
Is it possible (or even probable) that the only period of time during which humanity will know the truth of everything (or close to everything) will be mere moments before the end of our existence?

Can I Get Your Number?

Almost everything in the universe can be discussed in terms of numbers.

Atoms

Compounds

Chemical interactions

Temperatures

Velocities

Forces

Mass

Time

All of these things can be discussed in utter detail using the *same number system*. Meaning, numbers are a unifying language for much of science.

Does this mean numbers are real?

As in actual physical things, and not just a concept?

Choose Your Own Adventure

From the following, choose which one you *hope* is what happens after your death:

- *People who exhibit sufficient good in life go to a heaven of some sort, while everyone else suffers a worse fate*

- *When we die, all that we are is forever lost*

- *Reincarnation; either as a human again or a different animal type*

- *We ascend to some higher form of consciousness, meaning we're no longer human, but we retain some of what we once were*

- *We roam as spirits either forever or for a period of time*

And now choose which one you believe is *most probably* the truth.

Nihil

We've talked about nihilism before.

Nihilism - Life is without objective meaning, purpose, or value

But can someone truly be a complete nihilist?

Wouldn't someone with such a worldview immediately commit suicide, being utterly hopeless as they are?

For that matter, considering the questioning nature of the human mind, can anyone believe in any *'ism'* 100% of the time, no doubting ever?

A Big Limb to Go Out On

For purposes of this question, let us suppose that morality indeed exists. That good and evil are definable things. And that humanity has a tendency, though not a mandate, to be good instead of bad.
With that in mind, are any of the following reflective of a tendency toward immorality?

Believing in something without objective proof

Teaching children to believe in things without objective proof

Denouncing other humans for their beliefs without objectively disproving those beliefs

Desiring segregation of people based on their beliefs

Desiring the conversion of other people to the belief system of your choosing

When We Were Evil

Throughout most (though not *all*) of the modern civilized world, codes of morality have been put into place that are largely in agreement with each other.
Generally speaking, assault, theft, rape, and murder are among the actions viewed as immoral, and thus are illegal.
And yet...
Portions of societies have existed (and still do exist) in which these actions *aren't* viewed as immoral. In other words, they're perfectly acceptable to do in certain contexts.
Now imagine the entire world follows this reverse code. Assault, theft, rap, and murder are allowable.
Will human nature, believed by many to be generally good, eventually overcome this system?

Or are humans capable of remaining 'evil' indefinitely?

Product of the Past

Since we are, *none* of us, responsible for our own presence in this world, meaning that none of us created ourselves or willed ourselves into existence, does that reduce any of our personal responsibility in this life?
In other words, *every* human alive was given life without his or her consent. We didn't ask for this particular existence, and in fact, if given a choice, many humans might have chosen a different existence altogether.
Does not having chosen this life allow for a certain *moral flexibility*?
Or...
Must we accept a moral responsibility whether or not we asked for it?
And if so, why?

It Makes For a Great Party Favor

What role does *love* play in the universe?

Meaning, is it a chemically-induced emotional state designed to help living things procreate, protect our younglings, and generally not slaughter each other whenever we feel like it?

If you believe it's deeper than that, and that it has some meaning beyond chemical compounds floating through our blood, here's your chance to explain.

Seriously.

Go for it.

Forsaken

In the U.S., approximately 63% of people maintain a firm belief in a god. And of the remaining unaffiliated people, 83% maintain a general belief in a divine creator.

In many countries worldwide, the numbers are much higher.

That's a *lot* of people. In fact, odds are *you* are one of the believers, whether firm or not so firm.

Now imagine tomorrow every single human on Earth were struck with undeniable, absolute evidence that there was in fact *no god at all*.

Consider that many cultures rely on the existence of a god or gods for their moral codes, traditions, and indeed, their very happiness.

With this belief shattered, would these cultures descend into chaos? Would morality go right out the window?

Or would most people reject the proof and continue to believe as they wish?

Everything for Nothing?

First, a definition:

Determinism - a theory stating that acts of free will, natural occurrences, and all social events are determined by preceding events or natural laws

Meaning that no matter what we do, how we do it, or when it happens, *determinism* states the end result will be the same.

Kinda morbid, right?

Now...

Considering the universe and all the vast intergalactic events happening out there in the void, do you believe anything humanity does now or in the future will ultimately make a difference in how the universe moves along with its existence?

If not, meaning if humanity makes no difference in the end, answer this:

What is our value?

On Your Knees, Protons

In terms of scientists, atheists, and non-religious people…

…is physics god?

Moribund

If physical death is the absolute end of a human's existence,

And if the Earth is destroyed before humanity is able to populate other worlds,

And if in a future state the universe contracts onto itself or renders itself otherwise inhospitable to all life,

...is there still genuine value in the human experience?

Negating This Book

Define the value (if any) in asking questions to which the most honest answer is, *"I don't know."*

If you enjoyed The Ultimate Get to Know Someone Quiz, please consider leaving an Amazon review.

Thank you.

☺

Special thanks to:

Michael and Steven Kristensen
The JC Lunchroom
The G Man
Thomas & Chelsea Silva
The KScope Gang

For more life & libations, read J Edward Neill's:

Reality is best Served with Red Wine

A WALK AROUND EVERYTHING WITH J EDWARD NEILL

J Edward's Books:

Coffee Table Philosophy:

Reality is Best Served with Red Wine

Life & Dark Liquor

444 Questions for the Universe

Big Shiny Red Buttons

101 Questions for Single People

101 Questions for Couples

101 Sex Questions

101 Reasons to Break Up

Fiction:

A Door Never Dreamed Of

Hollow Empire – Night of Knives

The Hecatomb

Eaters of the Light series:

Darkness Between the Stars

Shadow of Forever

Tyrants of the Dead trilogy:

Down the Dark Path

Dark Moon Daughter

Nether Kingdom

About the Author

J Edward Neill writes philosophy and fiction for adult audiences. He resides in North Georgia, where the summers are volcanic and winters don't exist. He has an extensive sword collection, a deep love of wine and scotch, and a chubby grey cat named Noodle.

He's really just a ghost.

He's here to haunt the earth for few more decades.

Shamble after J Edward on his websites:

TesseraGuild.com

DownTheDarkPath.com

Téssera

Printed in Great Britain
by Amazon